BY THE SAME AUTHOR:

Memories from Paris to Stanford:
Life, Particles and Politics

The Human Condition:
Reality, Science and History

ESSAYS

Gregory Loew:

REGENT PRESS
Berkeley, California
2025

Copyright © 2026 by Gregory A. Loew

[PAPERBACK]
ISBN 10: 1-58790-721-6
ISBN 13: 978-1-58790-721-0

[E-BOOK]
ISBN 10: 1-58790-722-4
ISBN 13: 978-1-58790-722-7

Library of Congress Catalog Number: 2026900487

Cover Illustration and Design by Linda Maki

MANUFACTURED IN THE U.S.A.
REGENT PRESS
Berkeley, California
www.regentpress.net

Contents

Introduction 7

Gregory Loew:
Interviewed by Jean Deken 9

Winston Churchill and the
The French Revolution 36

The Difference Between
Gladstone and Disraeli 44

My First Car 45

The Portrait of a Lady
by Henry James
140 Years Later 47

Global Warming Revisited
After COP29 57

The Plastics Pollution Problem . . . 73

My Friend and Housemate
Moe Lerner 80

My Friend Guy Benveniste
Part 1: His Short Biography 83

Part Two:
My Celebration of Guy's Life 87

Memories of My Friend
Richard Pantell   90

The Significance of
the 1967 Line   94

Peace Between Israel
and the Palestinian Authority
is the Only Solution   98

It is Time to Begin Reforming
the U.N. Security Council 101

The Essence of the U.S.
Federal Social Security
System 1935-2025 107

Hallmarks of
Totalitarian States 111

FY2025 and FY2024
Federal Budgets 113

The Nuclear Weapons Age
at 80 [1945-2025] 115

Adolescence, Puberty
and Adulthood 120

Why Human Colonies on Mars
Are a Hopeless Project 122

Souvenirs d'Evian, Août 1939 125

Au Clair de la Lune 133

Introduction

This book is a collection of essays I wrote over the course of the last few years. The subjects are diverse and cover issues that concern me as a human being trained as a scientist. The titles of the essays are listed in the table of contents in the somewhat arbitrary order in which they are presented.

Readers that do not know me will learn details about my international education, and the first few years of my career at the Stanford Linear Accelerator Center (SLAC), in the first piece which is an interview of me in 2019 with Jean Deken, SLAC archivist for many years. The readers may also want to look at the two previous books I wrote since I retired from my fifty-year career at SLAC as a physicist. The titles of the books are *The Human Condition, Reality, Science and History* and *Memories from Paris to Stanford, Life, Particles and Politics.*

I grew up in Paris and in Argentina and have lived in the United States for the last 73 years.

For the fun of it, the last two essays in this book are in French. Bonne chance!

Gregory Loew: Interviewed by Jean Deken

JEAN DEKEN: It's Friday, November 15th, 2019, and I'm speaking with Gregory Loew:. Greg, I'd like to start with your early life. Where were you born?

GREGORY LOEW: I was born in Vienna, Austria, but never lived there. It was just an accident of my family's history. My father was from Bucharest, Romania and my mother was from Hamburg, Germany. We first lived in Bucharest, but a couple of years later we moved to Paris, France and that's where I grew up until I was 9 years old.

DEKEN: And what did your parents do? What did your father do?

LOEW: My father was a businessman, and my mother took care of me, my sister and my brother. She was very active at home, teaching us, helping us with schoolwork, taking care of our apartment or house but she never worked for money.

DEKEN: Did you grow up speaking French?

LOEW: Yes, I spoke some German because at first my parents were speaking German with each other, but as soon as we moved to Paris I learned French and from then on my first language was French.

DEKEN: And what kind of education did your parents have?

LOEW: They both finished high school. My father went to a small business school in Vienna. It was what he called "a school for commerce." He was also an accomplished pianist. My mother was very interested in art, water coloring and dance, and spent some time studying languages in a school in Geneva,

Switzerland. They were both intellectual and curious, they loved music, opera, theater and movies, they read a lot of books, but they did not have what we call a college education per se today.

DEKEN: And when you started school, what kind of school did you go to?

LOEW: That was an interesting thing. My parents loved to travel, and so I was signed up to go to a school in Paris called Le Cours Hattemer. It was a school where you went once a week for two or three hours. You were accompanied by your parents who sat in the back of the classroom with all the other parents. And there was a teacher who would run the class. She would examine you and you would get a chance to speak and to write and so on. And then you would go home with a pile of homework for the next week. If you were on travel with your family, the homework would be sent to you by mail. And for the first five years of my childhood, I would study at home with my mother. She would help me and supervise my homework, and then, the next week, we would go back to the school, take the tests, and so on. That was until I was nine.

Well, after that, when WW2 started, we moved to Buenos Aires, Argentina. And from then on, I went to a regular school, which was the French Lycee in Buenos Aires. There you learned everything in Spanish in the morning, and everything in French in the afternoon up to 6th grade. Four hours in the morning and four hours in the afternoon. After that, you could choose the French program or the Argentine program full time, but not both.

DEKEN: And were there any teachers that you had in school that were especially memorable or influential?

LOEW: Yes, I had several wonderful teachers. I had a science teacher named

Madame Renée Baccarat. She was a French biologist and had a great influence on me. Then I had a French mathematics teacher for two or three years in a row. His name was Monsieur Yvon Corbineau. He was also a wonderful teacher and I learned a lot from him. Those two people plus several others had an influence on me in two ways. One, they interested me in what they taught, and two, they showed me that I could be pretty good in math and science. And so, they had a marked effect on my subsequent career.

DEKEN: So, did they, or did your parents, encourage you or expect you to go on to college?

LOEW: Both. But there was never any question in my parents' mind that I would not go to college.

DEKEN: Even though your parents had not gone to college?

LOEW: Absolutely, they were totally set on sending me to university. The only question was, what would I study?

DEKEN: And was this the same for your siblings, that they expected them to go to college?

LOEW: Yes. Absolutely.

DEKEN: And so when you went to college, at what point did you decide to major in physics, or did you major in physics?

LOEW: My university career was complicated by a number of circumstances. When I finished the two successive French baccalaureates in Buenos Aires in 1946 and 1947 (degrees officially sanctioned by the French Government), I had to decide where I would go to college, and there was a conflict. One was that my parents thought that it would be good if I first got a more general humanistic education in France and then came to the United States to become more specialized in science and technology. The world situation with the Cold War and the Berlin Airlift in Europe was very bad at the time, so my parents were worried about that plan, but eventually it worked out. I went for three years to the University of Paris, the Sorbonne, and there I got what's called a "Licence es Sciences," in mathematics, physics, chemistry and electronics. These four subjects were studied fulltime, one year at a time.

Then, after that, I came to the United States as per the original plan. I was admitted to Caltech in the Electrical Engineering Department to study electrical engineering, electronics and physics. Richard Feynman's lectures on Quantum Mechanics were the best show in town, even though they were a bit too advanced for me to take for credit. After I got my M.S. Degree from Caltech in two years, at the encouragement of a professor who had just moved from Stanford to Caltech I applied and I was admitted to the EE Department at Stanford.

DEKEN: Who was the professor at Caltech

who encouraged you?

LOEW: It was Lester Field from Stanford who had gotten a tenured position at Caltech, with a lucrative consultantship at Hughes Aircraft. He was one of the inventors of the traveling-wave tube, which was a microwave device which worked very much like an accelerator in reverse: beam in, microwaves out! I took his class and he saw that I liked it very much. And he said, 'I think you will be happy at Stanford. If you want to get a Ph.D., go to Stanford.' So that's what I did.

DEKEN: So, when you came to Stanford, what year was that?

LOEW: That was in 1954.

DEKEN: Who was your thesis advisor?

LOEW: My thesis advisor at Stanford was Karl Spangenberg, a well-known vacuum tube specialist, also later involved with transistors. He was my supervisor, and Frederick Terman, famous for his book on Radio-engineering, was his boss. When I came to Stanford, all those people were still very accessible. So, I met Frederick Terman many times. He was a bookworm and could often be seen walking across the Quad reading a book! Because of his many accomplishments, Fred Terman is often considered the father of Silicon Valley.

In EE, we had a very close connection with physics and so I took several physics classes. My PhD degree was half with EE and half with physics courses. The thesis was on a special traveling-wave tube, which was a microwave generator. I worked with Karl Spangenberg and I also worked with a very sharp colleague of mine, C.T. Sah, who later became a very well-known specialist in transistors. When I was finishing my PhD, the transistor began to compete with vacuum tubes,

DEKEN: And so, as you are finishing up your PhD, are you looking around to decide what you are going to do next?

LOEW: Absolutely. That was a hard question for me, I didn't know what I wanted to do next. So, I did several things. In those days, if you earned a degree in EE from Stanford, all doors were open to you everywhere. So, I was invited to interview with Bell Labs, IBM, University of Illinois, Westinghouse, Hewlett Packard and others, and I got offers from every one of these companies, but I still couldn't decide. And then I ran into one of my ex-professors at Stanford, Marvin Chodorow, who was very well known at the time for his contributions to the design of klystrons, and I told him about my dilemma, that I really liked the idea of university work, and I didn't quite know whether any of these other companies would be as exciting. And he said, "Let me see if I can find anything of interest to you here at Stanford." After a couple of days, he called me and said: "Look, I can't promise you a job for life at this place because we don't know what's going to happen, but over in the Hansen Labs (right next to where I was in the Electronics Research Lab), there's a new project which is proposing to build a big linear accelerator. And they're looking for people like you. There are only about three or four people working on this project right now, but you would have to take a risk because there's absolutely no guarantee that this project will ever be approved." This was in May of 1958, it turned out to be my job for fifty years!

"No guarantee whatsoever the project will ever be approved by Congress because it's a big deal but if you want to do something interesting for a year or two and get a temporary job there, there's a man who is looking for people like you. His name is Richard Neal. He's basically the technical head of this project at the Hansen Labs, and he's willing to interview you, and if you guys can get together, you might get a job there. Again, no guarantee, but it should be interesting." So, I went to see Richard Neal and we hit it off immediately. He was a wonderful man, a very supportive person, as a matter of fact, I worked with him for my first 25 years at SLAC. But anyway, in those days there was no "SLAC," it was called "Project M."

"M" stood for "Monster." The idea was that this machine, this accelerator if it ever got built, would be a monstrous affair, three kilometers long. I thought about it, got very excited by this work, and decided to go for it. When I started to work with Richard Neal, there were several other people in the Microwave Lab aside from him that were influential in helping me and supporting me: Ken Mallory and Bill Gallagher, both very knowledgeable about microwave measurements, and En Lung Chu, the accelerator theorist.

However, the main person I saw every day was Richard Neal. and that's how I started to work for what eventually became SLAC.

DEKEN: So you started out on campus working for Richard Neal. What were the main challenges that you were faced with in Project M?

LOEW: The first challenge for me was to design the heart of the accelerator, a structure made out of copper, supplied by microwave power to accelerate electrons to about 20 GeV. Four smaller linear accelerators had been built at Stanford so far. The first was the MARK I, which was the creation of William W. Hansen, the founder of this electron accelerator technology at Stanford. The second was the MARK II which was a short experimental two-section accelerator, and the third one was the MARK III. The MARK III accelerator was built in stages and eventually reached about

1GeV in electron energy. It was where Prof. Robert Hofstadter got his Nobel Prize for his work on electron scattering of numerous atomic nuclei and where Pief Panofsky did his pioneering experiments with pi mesons. All four linacs used cylindrical disk-loaded waveguides (arrays of pill-box resonating cavities) for their accelerating structures. Richard Neal had written his entire PhD thesis on the Mark III. I wil come back to the Mark IV later.

The first thing that Richard Neal asked me to do was to see if I could explore all kinds of other structures to see if any might be more efficient. After a few months of research, the answer turned out to be no: the disk-loaded waveguide was the best. But a question remained: Should the cavities in the array all have the same dimensions (so-called uniform structure) which would lead to an exponentially decaying electric field in each section, or might it be better to design a new structure tailored to produce a constant electric field (a constant-gradient structure) by gradually decreasing the dimensions of the cavities along its length. The constant field was more conservative in that it was less prone to electric breakdown. But the constant-gradient structure design was more daring because it had never been done before. Each ten-foot accelerator section would consist of 86 cavities including two end-couplers to feed microwave power in and out, but each of the 86 cavities would need to have different dimensions, tapered to achieve the constant field. So that's the task I took on. To do the job I was offered the help of a very meticulous technician/engineer of German origin called Otto Altenmueller and a couple of graduate students. Together we designed and figured out the dimensions of these cavities. There were no computers at the time and everything had to be done experimentally, cavity by cavity. After about two years we came up with the design that became the SLAC accelerator structure for the next 50 years. There's a picture of Otto Altenmueller doing a microwave measurement in the SLAC Two-Mile Accelerator book.

DEKEN: Oh, The Blue Book. Okay. Okay. Very good. So this had never been done before?

LOEW: No, this kind of structure had not been built before. The idea of the constant gradient was something that Dr. Neal had only explored theoretically.

DEKEN: When you were experimenting and modeling the structures, were you actually working in copper?

LOEW: Absolutely, copper was the only way to go at the time.

DEKEN: Do you remember how many different structures you tested and designs you tried before you hit upon the one that you thought would work?

LOEW: The way we did it was to divide the measurement into 13 sub measurements. We made 13 different stacks of cavities and figured out what kinds of apertures, diameters, and so on they would have. We worked with the Fabrication Department under a very capable mechanical engineer called Arnold Eldredge, with another microwave engineer, Richard Borghi and a whole army of mechanical engineers/fabricators, machinists, furnace designers, etc. After they produced a few prototype sections, I had to test them on the Mark IV linac in the Hansen Laboratory. This linac had been run before with standard structures and it was also being used for cancer electron radiation therapy trials. When our prototype structures were installed inside the vault, I spent many nights there testing them with two klystrons and looking at the beam at the end. There was only room for two prototype sections at a time. We tested probably four or six structures until we finally decided that the design was fine.

DEKEN: You said that there was cancer treatment going on, was that something you participated in at the time?

LOEW: No. The electron beam was tuned up by an operator and the radiation treatments were done entirely by MDs from the Stanford Medical School which was just being moved to the Campus from San Francisco. X-ray radiation treatment was already in use but direct radiation with electrons was a new idea which looked promising for certain eye and other diseases. I had nothing to do with that work.

Meanwhile, in 1959 our 1957 design for Project "M" had been proposed by President Eisenhower, but the Congress had not approved it because the price tag was over a hundred million dollars. In those days this was a huge request of money for something not well understood, a two-mile-long linear electron accelerator. It took Dr. Edward Ginzton (then our director) and his Deputy, Dr. Panofsky, many, many trips to Washington to convince the AEC, the Atomic Energy Commission, and eventually the Congress, to approve and fund the project. In those days, just as today, most senators were not scientists and they did not know what this whole project was about. I vividly remember that there was once a senator who asked Dr. Ginzton: "Well, Professor, what is all of this for, what do you want to discover with this accelerator?' And Dr. Ginzton looked at the senator and replied, 'Senator, if I had an answer to this, I wouldn't be here." Surprisingly, the senator accepted this answer It would be inconceivable today! He would be shown the door!

So eventually these many discussions in the AEC Joint Committee and in the Congress led to the approval of the project for a hundred and fourteen million dollars. And at the end of 1962, Stanford got the contract to build the project and our jobs became a little more secure.

DEKEN: So at that point, did you transition up to the SLAC campus from main Stanford campus? Or when did that happen?

LOEW: Before the SLAC campus was ready, we moved out of the Hansen Microwave Lab into two large temporary buildings, M1 and M2, near the football stadium. They contained just cubicles, no offices. And that's where Dr. Ginzton and Dr. Panofsky ran the project together, although Ginzton was still Director. Then a couple of things happened. When the project was approved, Professor Ginzton, who was a very charismatic person, ran into a big problem because he had also collaborated with the Varian brothers who were running their company in Palo Alto. The two brothers, Sigurd and Russell Varian, both died within a very short time, leaving nobody of their stature at the helm. At that point, Professor Ginzton was offered the job of running Varian Associates, which still exists today, and he decided that being an engineer and not a physicist like Pief Panofsky, he would be better off running this company than to staying at SLAC, So, he transferred over to Varian in Palo Alto and Panofsky became Director. Meanwhile, we were looking for the best site for SLAC. The preferred site at the time was mostly under the Stanford golf course and extended for about three miles between Page Mill and Alpine Road. Very few people remember this today.

However, this site had to be built mostly deep underground, which would not have turned out to be very practical. So, after some further studies, we zeroed in on the current Sandhill site, between Alpine Road and Whiskey Hill Road. I remember, when I was still at Stanford, I once came up to look at this site, all by myself in my car. I entered on foot. There was no road and no people, just some nice oak trees and a cow. I took a picture of the cow It may well be the first picture ever taken of the lab! I may still have this picture somewhere in my files. Shortly thereafter, the Sandhill site was selected because it would be less costly to build a machine not entirely underground.

The first building that was built for people coming up from the campus was the Test Lab. The Test Lab was the place where we were going to do our microwave experiments and also where the klystrons were going to be built. Both myself and Dr. Lebacqz were the first to come up!

ESSAYS

LOEW: It was a very momentous epoch in international physics because in September of 1963, the International Accelerator Conference was scheduled to be held in Dubna, USSR. This was the first such meeting to be hosted by the Russians. I was invited with Drs. Neal, Panofsky, Matt Sands, Richard Taylor and a few others to attend the conference. This trip was an incredible occasion, both scientifically and socially. because for all of us it was the first time that we went to Russia and interacted with Russian people and scientists under the Communist regime. Also, it was a special time in my career because, along with my experience in designing accelerator structures, I had designed and built one of the first microwave particle deflectors that was capable of pushing charged particles sideways instead of forward. The Russians had built a device with the same purpose but much more cumbersome than mine, and my presentation was a great success.

DEKEN: It was called a what separator? What was its purpose?

LOEW: A "particle" separator. The idea was that if you could synchronize the radiofrequency wave in the deflector with a particle beam, you could separate some desirable particles from others at different speeds, that were undesirable. I built this device, again with the help of Otto Altenmueller. By 1966-1967, we installed four of these structures for different groups in the SLAC end-stations, to separate K-mesons and protons from electrons. They still exist today and 50 years later, one of them is used as a deflector for the LCLS!

DEKEN: So, once you're settled into the test lab, you are in your second-floor office, the one that Laura [O'Hara] and I helped you close down [30 years later]. But, once you're moved into that office and your group is with you now,

ESSAYS

what is your top priority?

LOEW: Once I knew and had let Dick Neal know that I had solved the constant-gradient structure problem, the next challenges were that in order to make a linear accelerator work, you had to have something to drive it. It was pretty much established at that time that the way to drive a linear accelerator with sufficient microwave power, you needed a klystron. A klystron can be made into a high-power microwave tube, and these tubes in those days were also designed in the Microwave Lab by a Dr. Jean Lebacqz. By 1959, the project that was being planned needed about 240 klystrons. This array of klystron tubes would be installed in a gallery above the accelerator and connected to the sections below by rigid copper waveguides. OK, so now the next question was how were we going to drive and phase this whole array? And since by that time I was pretty much done with the design of the accelerator itself, I was asked to design these next systems called the drive and phasing systems. If you look at the so-called SLAC Blue Book mentioned earlier, the next chapters after the accelerator design and fabrication have to do with the drive and phasing of these klystrons. This was a good job for me because it involved technology with which I was already familiar. To proceed I had to hire around ten new people. Our group was at first called the Microwave Department., Four or five engineers helped me with the drive system to provide the microwave power to drive the 240 klystrons and another four people helped me design the phasing system which had to make sure each of the 240 klystrons would be phased so it would contribute the maximum energy to the beam. This system involved a lot of modern electronics. Shortly thereafter, I was also asked to incorporate the Injector Group led by Roger Miller into my department, as well as the Instrumentation and Control Group under Ken Mallory. My department then became the Accelerator Physics Department, approaching 50 people. Subsequent details of my life and career, and the history of SLAC are illustrated in my book *Memories from Paris to Stanford.*

Winston Churchill and the French Revolution

When Winston Churchill was asked what he thought about the French Revolution, he answered: "It's too early to tell." I can only guess why he gave this whimsical and non-committal response. Was he being serious in the sense that we still need more time and perspective to express an opinion on the pros and cons of this historical event? What more is there to know?

Let us start by reviewing known facts.

A dramatic historical development such as the French Revolution went through several phases and had multiple causes, some of them distant and some more immediate. One of them was the effect of the French Enlightenment of the 18th century on public opinion. Towering over the Enlightenment was Baron de Montesquieu whose writings reached deeply into not only French society but also North American intellectuals, setting the stage for events to come. Montesquieu was a pioneer in the way he analyzed political systems from the Roman Republic to his time. He thought that a republic was characterized by the principle of virtue, monarchy by honor, and despotism by fear. He wrote in his Spirit of Law that countries would ideally be governed by three separate powers, the executive, the legislative and the judicial. He had a great influence on the founders of the American Revolution, particularly on James Madison.

Another cause of the French Revolution was the near bankruptcy of the monarchy inflicted by King Louis XV through his endless wars and abuses. His successor, Louis XVI, for the first fifteen years of his reign (1774-1789) ruled as an absolute monarch, although he had an indecisive personality. He tried to replenish the treasury's coffers by increasing taxes but was mostly unsuccessful. His Austrian wife, Marie Antoinette, was unpopular because of various scandals. In 1776, he allied France with the American Colonies against Great Britain, a move that prompted General Lafayette to join the American War of Independence but again cost a significant amount of money to his monarchy.

Although Louis XVI was generally anxious to be loved by his people, he made the mistake of deregulating the grain market, as a result of which bread prices rose abruptly and caused serious famines, starting in 1785. Louis XVI became increasingly unpopular and had to convene the Estates General in early 1789. These consisted of three groups, the Nobility, the Clergy, and the Third Estate (the bourgeoisie and the poorer classes) which had little power. The bourgeoisie paid most of the taxes.

On July 14th, 1789, the people of Paris stormed the hated Bastille prisons. This event marked the beginning of the revolution. The King was forced to recognize the legislative authority of the National Assembly. In August 1789, this Assembly was responsible for publishing the Declaration of the Rights of Man and of the Citizens, drafted initially by Lafayette, assisted by Thomas Jefferson. The document had a worldwide impact, giving all French men (not women) equal social and political rights, freedom of expression and freedom of religion. Women did not get the full right to vote until 1945.

Had Louis XVI been more skilled at this point, he might have accepted and enacted a Constitutional Monarchy like Britain. Instead, he tried to flee Paris in 1791 with his entire family, but was caught in Varennes and brought back to the capital. At the time of the insurrection of 1792, the monarchy was abolished, and the First French Republic was proclaimed on September 21st, 1792. The king was demoted, accused, and convicted of treason, and on January 21st, 1793, executed on the guillotine. Marie Antoinette was jailed and guillotined six months later.

Two of the most famous and rival politicians who had risen to power in the government of the *Convention Nationale* and the ensuing *Terreur regime* were Georges Danton and Maximilien Robespierre. Both successively wielded considerable influence in opposite political positions. Both ended badly and their historical legacy is still controversial. Danton was tried for alleged corruption and royalist sympathies, and guillotined. By 1795, Robespierre was accused of seeking dictatorial power and, without trial, was also guillotined.

The numbers of civilian victims of the French Revolution were about 16,000 guillotined (80% of them commoners), 20,000 shot, stabbed, or drowned, and about 170,000 in the civil war of the Chouans in the Vendee.

The Marseillaise, composed by Claude Rouget de Lisle in Alsace, symbolized France as a patriotic national power. It became the battle cry of the French Army and eventually the French National Anthem. Ironically, since most people in Alsace did not speak French, a version with German lyrics was published there.

The First French Republic went through chaotic times and was unable to govern France effectively. Napoleon Bonaparte, a national hero but an authoritarian at heart, put an end to it in two steps, first through a coup (18 de Brumaire) in 1799 by which he became ruling First Consul, and then when he proclaimed himself as Emperor of France in 1804. He legalized the metric system, replaced the Marseillaise by a different hymn, enacted the Napoleonic Civil Code, and was a military genius. Unfortunately, he fought wars all over Europe against seven successive coalitions of monarchies, the last of which defeated him at Waterloo in 1815. These wars were probably responsible for the death of 1 to 2 million soldiers.

This was a long and complicated story, but it was not too early to tell. It is unclear what Churchill could have told us is because the French Revolution had endless consequences all over the world. History is an infinite chain of interlocking events. But unlike a play or a movie, it is not written ahead of time. It just runs its course from cause to effect. We would like to think that it runs toward progress, but what exactly is progress? Decreased human suffering and more universal joy may be succinct answers. By these criteria, the French Enlightenment of the 18th century and the Declaration of the Rights of Man and of the Citizens were progress, but the bloodshed of the Revolution and the Napoleonic wars were not. Did this Revolution usher the way to make France easier to govern? Not really since after 1815, the country lived through three more monarchies, one empire, and four more republics. Even the Fifth Republic, with all its improvements, still has to deal with popular movements emerging *on the street,* and insoluble political tensions between left, center and right, especially if the President and Prime Minister belong to opposite parties, which occasionally happens.

The Difference Between Gladstone and Disraeli

Queen Victoria, during her long reign (1837-1901), had more than 33 Prime Ministers. Two of the most famous were William Gladstone and Benjamin Disraeli.

A young lady who was taken out for dinner by Gladstone one evening, and by Disraeli the subsequent evening, related her impressions as follows:

"When I left the dining room after sitting next to Mr. Gladstone, I thought he was the cleverest man in England. But after sitting next to Mr. Disraeli, I thought I was the cleverest woman in England."

My First Car

I bought my first car, a 1938 Oldsmobile, from a used-car dealer on Colorado St. in Pasadena when I was at Caltech. I paid $200 which I had saved, giving private French lessons. I should have known better, but a day later, going down a slope, the drive shaft which had been welded illegally broke, and the car had to be towed back to the dealer. It took a huge shouting match for him to admit his dishonesty and to fix it. The Olds was a large limousine and a gas guzzler, but gas was ten cents/gallon, and I did not care. There was hardly any public transportation in Los Angeles, and it changed my life.

GREGORY A. LOEW

In 1954 when I moved to Stanford, I drove my Olds loaded with all my belongings to the Bay Area. It lasted another year but eventually gave up the ghost on the way up Marin Street in Berkeley. The slope was too much for the old lady!

The Portrait of a Lady by Henry James, 140 Years Later

This essay is dedicated to the memory of my good friend Anne Knight of Palo Alto, California, who died in 2020 after a long battle with lung cancer. Anne had been an English major at Harvard, she was a voracious reader and a talented critic. Henry James was one of her favorite authors. Anne recommended I read this book about twenty years ago.

"*The Portrait of a Lady*" is one of James' earlier novels (1880) augmented by his later Preface (1907). In the version I read (The Modern Library, NY), it is introduced by Anita Brookner. I would characterize it as a pre-Freudian psychological novel with exquisite dialogues and descriptions of situations and places. It moves slowly but steadily.

Isabel Archer, whom James calls his heroine, is a woman in her early twenties, from Albany, NY. Her father, a good hearted, happy go lucky gambler and avid traveler, has recently died and Isabel and her two married sisters are now left without parents. Isabel's aunt, Lydia Touchett, who resides mostly in Florence, visits Isabel in Albany, takes an interest in her and invites her on a grand tour of Europe. The story starts in England in an idyllic country house, 40 miles west of London, by the name of Gardencourt. Mr. Touchett, an elderly American, lives there with his loving son, Ralph. He is essentially separated from Mrs. Touchett, they tolerate each other at most for one month per year. Mr. Touchett settled in England many years ago and made a fortune in London as a banker. We meet him having his ritualistic afternoon tea in the garden with Ralph and his friend Lord Warburton, as Mrs. Touchett arrives with Isabel from the US.

James tells us in the Preface that his novels, certainly this one, are constructed much as his Paris friend, Turgenev, described to him the gestation of his own novels. They are all based on real people who are "disponible," i.e. known and available to the artist, whom he or she transforms and around whom they weave the story. This source of inspiration is unlike that for Nineteenth Century French novelists who start from the other end, namely with a grand and overarching architecture in which characters find their place. James's characters, one learns in the notes, are inspired by friends and relatives (Isabel may have been modeled after one of his cousins) or even by characters from other novelists like George Eliot. All these people are contemporaries, indeed Isabel is probably only ten years younger than James (born in 1843) since in her childhood she was already aware of the on-going Civil War. Just like the author, all the Americans in the story are very much attracted to European cities, especially London, Paris, Rome, Florence, and Venice[1] where the book was written. It is interesting that indeed in those days many Americans who could afford it, thought of emigrating to Europe. In James' words, many adapted there yet they never tried to assimilate completely. He himself died in London, naturalized an honorary British subject!

What is the novel's theme? We are told that it is the story of a "certain young woman confronting her destiny." Nice impressionistic words, but what do they really mean? How can we confront something that does not really exist? Destiny is realized ex-post facto, unless we think of it in very vague terms: a set of potentials stemming from early conditions of nurture and nature. Only a novelist, playing God with his imagination, can predestine a character. I have no way of proving this but I suspect James didn't know when he started, where and how Isabel would end her odyssey, particularly since he doesn't end it. He wanted to paint her as a free spirit and let her fly. To observe her, he placed four suitors in her way: the monolithic and lovesick Boston cotton mill entrepreneur, Caspar Goodwood, the noble and immensely rich Englishman, Lord Warburton, her consumptive cousin Ralph Touchett, and the manic Florence-bound art lover, Gilbert Osmond. A few other indispensable characters were added to the plot: the enigmatic Madame Merle, peripatetic European traveler originally from Brooklyn, Osmond's doll-like young daughter Pansy, brought up in a convent and programmed to love and respect her father unconditionally, and Isabel's American reporter friend, Henrietta Stackpole who in the course of her European forays finds her British companion, Mr. Bantling. Two more individuals are needed, near the denouement, Ned

Rosier, Pansy's admirer, and Countess Gemini, Osmond's sister, married to an insignificant Florentine Count. Between Gardencourt, London, Paris, Florence and Rome, all these people keep appearing and reappearing at expected and unexpected times, holding the suspense.

As I stated at the beginning, the story, albeit long, is exquisitely told, the dialogues are totally believable, and no movie or photographs of the settings could do justice to the richness with which they are described. One can appreciate, 140 years later, how serialization of the novel in the Atlantic Monthly (US) and in MacMillan's Magazine (UK) would keep their readers captivated and on their toes for months.

What distinguishes "The Portrait" from a contemporary novel and makes it interesting is what is said and what is not. I indicated earlier that this is a pre-Freudian psychological novel. By this I mean that we are told what people think, we hear them talking beautifully to each other (even when they get rather outspoken and tough) and see them moving about in their environment. Except for Henrietta Stackpole who is seen at work in the novel, all the other characters are self-sufficient, rich or supported. They move, unimpeded by financial constraints, in the psychological cauldron. But without the benefit of their unconscious drives and the web of their genetic makeup, we have to take their actions at face value. Isabel is charming, beautiful, smart, curious and independent, until she finally falls into a trap. One doesn't quite understand why she becomes so vulnerable to betrayal after being so insightful, strong and free. By existentialist standards she fails miserably. Gilbert Osmond is the least understandable, he seems to fall for Isabel, which then makes it hard to see why he ends up hating her so much. He reveals himself as a sophisticated art collector but a conformist automaton. Why?

Isabel is not that much of a threat to his security. Madame Merle, with today's insight into genetics, is the true "selfish gene" in operation, looking for assured success to propagate her genetic make-up.

The novel doesn't depict crowds, poverty, common people. There is of course no mention of sex except by innuendo in two or three places, one crucial to the story. Money is discussed openly, we are told that seventy thousand pounds can keep somebody going forever, but forty thousand francs is not much. In one description, James talks about "atoms of light," the image is beautiful but inevitably dated since nobody believes anymore that light consists of atoms. What hasn't changed since 1880 is the vulnerability of marriage: the enterprise was just as precarious then as today. Only divorce didn't seem as obviously available as an outlet. But the result was the same. I think one of the strong reasons why young people should read this book today is because of this pervasive problem, the ever unpredictable risk of committing to such a contract. People fell in love then like they do now. However, the frequently appearing expression "of making love" meant something quite different then: it just referred to the act of intensely expressing this feeling of love to another person, not necessarily doing anything!

As remarked by others, Isabel and her admiring cousin Ralph are the only "moral purists" in the novel. When Ralph dies, reason seems to die with him. He realizes that the generous secret gift he bestowed on Isabel before his father's death may actually have been coveted by Osmond and led to her demise.

Perhaps, as Anita Brookner suggests, James doesn't really reveal Isabel's end because he wants us, the readers in our imagination, to let her find her ultimate destiny. But how can we do this? How can we construct a better future and denouement for such a thoughtful and rational woman after she falls into such an impossible marriage? Leading her to a tragic end like Madame Bovary or Anna Karenina is not an option. But associating her to a better man than Gilbert Osmond without today's obvious solution of divorcing him doesn't seem feasible in the context. So what is our exit for Lady Isabel?

(1) James's description of where he wrote most of the book in Venice was of special fascination to me. It turns out that by coincidence, several years ago, I stayed with my family in the same passage to San Zaccaria Church perpendicular to Riva degli Schiavoni. It is now part of a hotel "dependance".

Global Warming Revisited After COP29/30

This article is an updated version of a chapter I published in 2023 in my book on *Memories from Paris to Stanford*. It is written after Hurricane Helene, President Trump's re-election and the COP29/30 meetings in Azerbaijan (November 11-22, 2024) and Belem, Brazil (November 10-21, 2025. The article reviews the causes of global warming, which greenhouse gases and human activities are most responsible for it, and what may lie ahead. The year 2024 turned out to be the hottest on record with 57 Gigatons of greenhouse gases released in the atmosphere.

GREGORY A. LOEW

How the greenhouse gases work

To summarize the subject once again, global warming is caused by the following mechanism:

The radiation from the sun heats up our earth. This heat would normally be re-radiated into outer space by infrared radiation. However, our human activities annually emit more than 50 billion tons or Gigatons of greenhouse gases of equivalent CO_2 (carbon dioxide). Of these, about 30-50% are absorbed by the oceans (thereby acidifying them and seriously affecting their entire ecology), and about a quarter are stored in trees and plants worldwide through the process of photosynthesis via chlorophyll in the leaves. The remaining ~20 Gigatons accumulate each year in the atmosphere, and similarly to the glass covering a greenhouse, block the infrared radiation from escaping by reflecting it back to earth and raising its temperature. The total accumulation of these gases is about 3200 Gigatons. Since the beginning of the industrial age, until 2022 it had produced a temperature increase of about 1.2 degree Celsius with a total equivalent CO_2 gas density by volume of about 450 parts per million molecules in our atmosphere. The reason the word "equivalent" is used is because ~70% of the effect is due to CO_2. The remaining 30% is caused by CH_4 (methane in natural gas) which has a Green House Potential (GHP) ~28 times greater than CO_2, smaller amounts of N_2O (nitrous oxide) with a GHP of 280, as well as fluorine and other greenhouse molecules.

Future temperature increases

In 2015 when the Paris Agreement on Climate Change was signed by 195 countries, it was projected that if the world wanted to avoid the worst and most disruptive effects, the total temperature increases by 2050 would have to be limited to 1.5 degree Celsius. Unfortunately, in 2023 it has already reached 1.48 degrees, and the planet is witnessing serious storms, hurricanes, floods, droughts, wildfires, and sea level rises. As examples, one third of Pakistan, Greece and Libya were flooded in 2022-2023, and hurricanes are devastating the U.S. and Puerto Rico. Unless the countries that produce the most greenhouse gases make draconian cuts soon, we appear to be headed towards a 2.7 degrees Celsius rise or more by 2050, which could be disastrous. At these higher temperatures, ~30% of the high-altitude glaciers could melt, the Northern polar ice cap could disappear entirely, and the Siberian tundra permafrost may warm up and release its huge amount of methane, three phenomena that will be impossible to reverse.

Major responsible countries

When we look at all the countries in the world, we see that a group of six are responsible for just under half the annual total release of greenhouse gases, as shown in *Table 1* (*right*).

The sum of emission contributions of these countries was ~26.7 GT (53% of total) and the GDP $68.4 Trillion (about 69% of world $100 Tillion GDP) in 2022.

One book on global warming that made a strong impression on me was *How to Avoid a Climate Disaster* by Bill Gates. The book did not have all the answers, but it created a complete inventory of all the human sources of greenhouse gas emissions and where technological innovations, and investments are needed to curtail them. *Table 2* (*next page*) summarizes the percentages contributed by each human activity in the world, to which I have added the percentages of emissions

Table 1: Largest country emissions and GDP statistics

Countries in 2022	Giga-tons	Tons/Capita	% of Global Emissions	GDP in $Trillions
China	14	10	28%	18.3
United States	4	12	8%	23
European Union	3	7	6%	17
India	3	2.3	6%	3.5
Russia	1.7	1.1	3.4%	2.1
Japan	1.0	0.8	2.0%	4.5

in the U.S. for comparison.

With this tabulation, Gates identified what he calls the Green Premium or excess percent cost incurred by a green technology that would avoid the release of greenhouse gases and figured out systematically what innovations are needed as early as possible to replace fossil fuels by the most efficient fully green sources. Gates addresses these two challenges in lengthy detail which I cannot possibly match here, but I am summarizing some of them below, adding some ideas of my own.

Table 2

Human activity	% of world emissions	% of U.S. emissions
Producing electricity	27%	27%
Making things (steel, cement, plastics, etc.)	31%	22%
Growing things (plants, animals), land management	19%	10%
Transportation (cars, trucks, ships and planes)	16%	28%
Keeping warm and cool (heating, cooling buildings)	7%	12%

ESSAYS

Needed technical innovations

1) **Electricity:** Many of the activities listed in Table 2 above include the use of electricity. If we consider that we will have to charge our fully electric car fleet and the upcoming increase in world population, the need for electricity may as much as triple (currently 5000 gigawatts of power). All of this will have to be clean green electricity. It must be generated by renewable solar or wind energy (which are both intermittent), some more hydroelectricity, possibly nuclear fission reactors (if they can ever be made safe, and their radioactive waste sequestered for thousands of years), and possibly nuclear fusion reactors. The very recent breakthrough at LLNL's National Fusion Ignition Facility is good news although it only yielded 1 kw-hour out for ½ kw-hour in. Barring any surprises, a commercial electric power station based on fusion is still ~30 years away. It will not create as much radioactive waste as fission reactors,

but it will still not be free of it.

The intermittency problem (daily and seasonal) of solar and wind may not be solved entirely by storage batteries alone, but the energy may be stored in the future in hydrogen generated by electrolysis of water with excess renewable electricity. Safe hydrogen storage still requires considerable R&D.

2) Manufacturing and construction: The green premium of making steel and plastics may be brought down close to 1 by replacing heat from greenhouse gases with electrical heat. Plastics, of course, have the additional problem that most of them are not recyclable and, when discarded, pollutes land and oceans. This is a very serious worldwide problem that I address in a separate article.

The making of cement from limestone calcium and sand does not yet seem to have an affordable green alternative. A revolutionary invention is urgently needed for the construction of buildings, dams on rivers for hydroelectricity, and bridges, let alone dikes to protect us from sea-level rise.

3) Agriculture: With the projected world population growth, at least 50% more food will be needed. Plant growth can be increased with fertilizers containing phosphorous, potassium and nitrogen, but nitrogen has a problem because it turns into nitrous oxide which is a potent greenhouse gas. We will need to waste less food and produce better fertilizers to promote regenerative agriculture.

Cattle and pigs, by belching and farting methane produce about 5% of world greenhouse gases. A recent discovery reduces the methane belched by cows by feeding them small amounts of seaweed. This would be good news if implemented on a large scale. However, growing cattle and pigs is a very inefficient method of providing us with proteins and fats. The best way would be to drastically cut down on our red meat consumption. This would also result in a corresponding decrease in the cruelty to animals when they are raised and butchered.

As far as land management is concerned, the most important step we can take is to completely stop the destruction of rain forests (already cut back by 17% in Brazil). On the other hand, planting a billion new trees in the world as some have suggested is a losing proposition. We wouldn't even find the necessary water!

4) Transportation: There are currently about one billion cars in the world, most of them propelled by gasoline. By ~2050 they will all have to be replaced by electric vehicles with inexpensive rechargeable batteries. The same is true for small trucks and buses. The good news is that these electric vehicles are significantly more energy-efficient than cars running on gasoline and cost less than half to drive for the same number of kilometers.

As to large trucks, ships and planes, batteries are too heavy for them, and we will have to produce biofuels or electro fuels to propel them. Sugar cane and switchgrass are good sources but much more R&D is needed in this area.

5) Buildings: Buildings will have to be much better insulated and heated and cooled entirely by electricity. The efficiency of current air conditioners can easily be doubled, but the use of gas for air conditioners and water heaters can best be replaced by heat pumps.

In all these areas, we must increase conservation and discourage wasteful consumption.

Are there any other methods available to 1) decrease greenhouse gases in the atmosphere, or 2) decrease the sunlight heating the earth? Yes, but they are long shots. The first category includes Carbon Capture and Storage (CCS) at the point of production like a cement factory, but the problem of where to store the gas securely still needs some work. Another possibility is to directly capture the gas from the atmosphere with an absorbing surface. This process works but may not be efficient enough. The second category includes reducing the sunlight heating the earth by releasing billions of light-reflecting sulfur micro pellets into the upper atmosphere. However, this form of geoengineering may take at least ten years to develop and a fleet of high-altitude airplanes to spread around the planet, and it may cause irreversible collateral damage. Another less invasive technique would be to make clouds brighter and more light-reflective by seeding them with salt.

What happened at COP29?

COP29 was attended by a record 85,000 people but its achievements were not as promising as COP28. The meeting was dominated by a very contentious debate between the richest countries most responsible for global warming and the much poorer developing countries. The latter asked the former for a subsidy of $1.3 Trillion per year to pay for their decarbonization, but in the end got only a commitment of $300 Billion per year, insufficient in their opinion.

Where are we at this point?

We have seen that to solve our global warming problem, we will need many innovations. If all these innovations had a green premium of less than 1, i.e., if the green technology were less costly than the current non-green one, we could count on the market economy to adopt it naturally within a short transition period. Unfortunately, this is not yet the case. To get to zero-net GHG emissions by 2050, governments must intervene at the national and local level. Examples of success stories are listed below.

China is the world leader in the production of both Electric Vehicles (EVs) and the batteries that power these EVs; though the two stories are not directly linked, both were caused by the coming together of favorable government policies, people with an entrepreneurial spirit and the persistence required to develop new technologies.

The success of solar power in India too can be attributed to the coming together of people, policies and technology to generate clean energy.

By 2022, Denmark was producing close to 20 Terawatt-hours of electricity from wind farms, about half of its total electric power.

In Mexico the surprising election of President Claudia Sheinbaum who earned her Ph. D at the Lawrence Berkeley National Laboratory in Energy Engineering (1994) and has been deeply involved in Mexican environmental policies in recent years, bodes well for the future influence of her country.

After President Obama promoted and joined the 2015 Paris Agreement, the Biden-Harris Administration with its 2022 Inflation Reduction Act has dedicated

$369 billion to deal with the problem domestically over the next seven years or $50 billion per year, compared with our defense budget of $800 billion per year!

Independently of governments, we as individual consumers have our own options. We can buy electric cars, which will be helpful as long as we recharge their batteries with clean green electricity, and we can install heat pumps outside our houses to get rid of natural gas. We can switch from cow to soy or oat milk (just as good), and to paraphrase Nancy Reagan, we can "just say no to red meat!" The cattle ranchers and butchers will not love us for this and will fight these steps tooth and nail, but they will have to adapt.

In San Mateo County where I live, the forward-looking Peninsula Clean Energy coalition already delivers fully renewable electricity to the entire county. We also have the very climate-proactive State Senator Josh Becker who passed more than twenty climate and solar-friendly environmental bills. Stanford University with the Precourt Institute, the Woods Institute for the Environment, and the Doerr School of Sustainability is also making important R&D investments in the field.

What lies ahead?

All the above initiatives constitute good news, but they will not be sufficient to get us to net-zero GHG emissions by 2050 with less than 1.5 degrees Celsius temperature increase to avoid a worldwide climate crisis.

It is most unfortunate that President Trump continues to call global warming a hoax and has again withdrawn the US from the Paris Agreement. In the short run, this is a bad step. But in the long run, this will probably not matter too much because the cost of renewables will prevail in the market. The recent impressive progress made by geothermal installations is most encouraging.

The Plastics Pollution Problem

The plastics pollution problem poses a worldwide threat to our entire planet, along with global warming. The problem is complicated by the fact that there are seven different categories of plastics (polyethylene terephthalate (PET), high-density polyethylene (HDPE), low-density polyethylene (LDPE), polypropylene (PP), polyvinyl chloride (PVC), polystyrene (PS), and others including polycarbonates), each with its different characteristics, ingredients, toxicities and degrees of recyclability. Some are not recyclable at all; some can only be recycled once. Their damage to the environment can be ascribed to **three general factors:** 1) they contribute to global warming because their manufacture requires using energy in the form of heat, 2) plastics are not biodegradable and in the cases they are recycled (currently less than 10% worldwide), this also requires heat or mechanical shredding energy, and 3) when dumped into landfills (~85% in the U.S.), they stay there for hundreds of years, and when dumped into waterways and oceans, plastics are absorbed as small pellets by fish, killing marine life, affecting biodiversity, getting into all our food chains, and thereby causing serious effects on human health.

This said, plastics are such an important part of our world economy that they can no longer be eliminated. They are everywhere: in all industrial, agricultural, military, scientific and medical equipment, in motorcycles, cars, boats and airplanes, in buildings, electrical, thermal and acoustic

ESSAYS

insulation, foams, solar panels, satellites, refrigerators, TVs, radios DVDs, CDs, computers, keyboards, iPads, phones, in light fixtures, furniture, carpets. pipes, tubing, gardening tools and machines, fishing nets and equipment, glues, paints, coatings, ropes, threads, clothing, backpacks, textiles, cleaning materials, brushes, lenses, decorations, plastic plants, most toys, chewing gum, pens, medical syringes, contraceptives and pills, kitchen utensils, bottles, urinals, suitcases, containers, bags, packaging and wrapping for food ubiquitous in supermarkets, and so on.

Bulk or "virgin" plastics are almost entirely made from petroleum products by large oil companies. They are transformed into all the above products because they are relatively cheap, durable, easily moldable into any shape, and lighter than glass or metals. Their durability is also their curse!

As of 2022, fifty nations had joined the *High Ambition Coalition to End*

Plastic Pollution, which was pushing for a treaty by the end of 2024 with binding constraints, targets and controls for every nation that signs the pact. The "ambitious" group of fifty nations grew to 170 in 2024. Below were some ideas for governments, manufacturers and consumers to consider:

1) Any energy to manufacture or to recycle plastics should be green electricity.

2) Recycling should be increased as much as possible, at least to 20%.

3) Given that 40% of plastics are single-time use, many of them should be replaced by biodegradable materials. This would include bags, cups, plates, bottles, wrapping, and more. This has already been enforced in some countries.

4) "Extended Product Responsibility" (EPR), which is commonly used in Europe to assign full life-cycle costs to products, is just beginning to appear

in the U.S. At least some additional sales taxes and fees should be created to cover the cost of plastic disposal.

5) Wood products should be given preference when possible.

6) The incineration of plastics, which pollutes the atmosphere, should be forbidden.

7) Dumping plastics in poor countries should be abolished. If you have any doubts, watch the video of the Rio Motagua and its adjacent beaches in Guatemala!

8) Dumping plastics into the environment, beaches, waterways, and oceans should be outlawed and severely fined.

9) Landfillls for plastic waste should be the last resort, not the first.

10) Much more research should be funded to see what can be done to reduce plastic waste and its toxicity.

GREGORY A. LOEW

As of September 23, 2024, it was encouraging that California Attorney General Rob Bonta announced the filing of a lawsuit against ExxonMobil for allegedly engaging in a decades-long campaign of deception that caused and exacerbated the global plastics pollution crisis. In a complaint filed in the San Francisco County Superior Court, the Department of Justice alleges that ExxonMobil has been deceiving Californians for half a century through misleading public statements and slick marketing promising that recycling would address the ever-increasing amount of plastic waste ExxonMobil produces. Through this lawsuit, the Attorney General seeks to compel ExxonMobil, which promotes and produces the largest number of polymers—essentially the building blocks used to make single-use plastic—that become plastic waste in California, to end its deceptive practices that threaten the environment and the public.

On a more pessimistic note, the

meeting of 170 nations taking place in Busan, South Korea, the last week of November 2024, that was supposed to agree on a final binding international treaty, failed. The idea supported by most countries to severely decrease plastic production (now close to half a billion tons per year) was blocked by the oil producers Russia, Iran and Saudi Arabia. The meeting was adjourned and resumed in Geneva in August 2025 where it was blocked again. The goal of putting an end to the problem entirely by 2040 is still not in sight.

Acknowledgements

I want to thank my granddaughter Avalon Surratt, a budding plant geneticist, who encouraged me to pursue this study, and my friend Ben Lenail who familiarized me with the Extended Product Responsibility concept described in point 4.

My Friend and housemate Moe Lerner

Moe Lerner was an undergraduate at Caltech when I met him. When we both came to Sanford in September 1954, he enrolled for an M.S. in Mechanical Engineering, and we decided to become housemates.

Moe's parents met and married in Moscow, USSR but emigrated to Mexico in the early 1920s. They had a daughter in Moscow, and Moe was born around 1932 in Mexico City. They moved to Los Angeles a little later.

Moe and I found a perfect little house in South Palo Alto in a mostly black neighborhood. The house belonged to a Swedish woman who rented us the house for $70/mo. under the condition that we watered her lawn and petunias with great care, which we did. Her name was hard to pronounce so we called her Mrs. Petunia.

Moe was an excellent student. He was very interested in women and sex, and he loved to sing arias from Lucia de Lammermoor by Donizetti.

We had great dancing parties with Stanford friends in the house, but Moe eventually latched on to a nice Mexican woman named Rita from San Jose. After he got his M.S., Moe moved back to Los Angeles with a good job at the Jet Propulsion Laboratory and lived with Rita for a few years. His parents, who were petrified because they wanted him to marry a Jewish girl, asked me to convince Moe to break up the relationship, which I refused to do. Eventually, they separated on their own and Moe met Barbara, indeed a Jewish girl.

Moe's parents arranged for a big wedding in a L.A. club to be officiated by a rabbi. Moe had to meet him ahead of time to discuss the service, at which point he said: "Rabbi, I have to confess to you that I don't believe in God" to which the rabbi answered

"Oh, don't worry about that, I don't either."

The ceremony, which I attended, was flawless.

Moe and Barbara had two daughters but, after many years, they divorced. Moe was becoming completely bald, and he bought himself some fancy wigs. He then met another Jewish woman and married her. After they had a son, he divorced her. Meanwhile, he had gone into real estate, designed, built and sold large apartment houses, and became very rich. He built himself a big, elegant house in Malibu where he moved in with another Mexican woman. At this point, we lost contact and our friendship ended.

My Friend Guy Benveniste
PART ONE
His Short Biography

Born in Paris, France on February 27, 1927, Guy Benveniste attended school at the Lycee Buffon in Paris, the Lycee Louis Barthou in Pau and the Lycee du Parc Imperial in Nice. In 1942, he moved to Mexico City and attended The American High School, graduating in June 1944. Admitted to Harvard College in 1944, he graduated with a BS in Engineering Sciences and Applied Physics in 1948. He obtained an MS in Mechanical Engineering in 1950.

From June 1950 to September 1952 he was a Construction Engineer at the Lecheria Power plant near Cautitlan in the

Estado de Mexico. He then joined the staff of the General Manager of the Mexican Light and Power Company where he undertook studies of future power demand and financial requirements for an eventual loan from the World Bank (IBRD).

In 1954, he moved to the US and joined the Stanford Research Institute (SRI) in Menlo Park, California. There he undertook studies on the economics of Solar Energy. Later he joined the International Division of SRI and undertook research on economic development in third world countries. During that period he published articles on Solar Energy and a study for the US senate Committee on Foreign Relations on "Possible Nonmilitary Scientific Developments and Their Potential Impact on Foreign Policy Problems of the United States" with Eugene Staley in 1959. Later a study on Africa was published: "Handbook of African Economic Development" with William E. Moran jr. Praeger 1662. In 1961 he was seconded to the Henry L.

ESSAYS

Labouisse Task Force on the reorganization of US foreign assistance leading to establishing the USAID.

In December 1961 he became Special Assistant to the Assistant Secretary of State for Educational and Cultural Affairs. He went to Spain to plan future cultural exchanges and attended the UNESCO Conference of Ministers of Education and Economic Planning in Santiago Chile in March 1962. In June 1962, he joined the staff of the World Bank to work on education. He went on mission to Afghanistan in the fall of 1962 resulting in a loan to that country in 1964. He was transferred by the Bank to Paris in 1963. He was instrumental in the creation of the International Institute of Educational Planning (IIEP) initially sponsored by UNESCO, the World Bank and the Ford Foundation. There he authored several papers on educational planning.

In June 1965, he returned to California to attend Stanford University where he received a PhD in Sociology of Planning in 1968. He was then appointed to the Graduate School of Education at the University of California at Berkeley. He remained on the active faculty until his retirement in 1993. During that period, he published numerous well-known books and papers on bureaucracy and the politics of planning. He was an early student of the process of planning and of its political elements. "The Politics of Expertise" was first published in 1972. His "Bureaucracy" was published in 1977. In his planning books, he argued that planners had to go beyond technical analysis to develop necessary coalitions to make plans a reality. His books on bureaucracy focused on the role of professionals to achieve organizational flexibility. His books were influential and widely used.

Guy Benveniste passed away peacefully on December 3, 2022 at the age of 95.

PART TWO
My Celebration of Guy's Life

I met Guy in 1955 at Stanford University during a weekly reunion of the International Club. We began to chat, and it didn't take long for us to realize that our trajectories in life had much in common. We had both been raised in Paris, had both immigrated to Latin America because of Hitler, he to Mexico and I to Argentina, and we had both come to the U.S. to study, he at Harvard and I, after getting an undergraduate degree in Paris after the war, at Caltech for my M.S. and Stanford for my Ph.D. Without any difficulty, we could communicate in French,

Spanish and English seamlessly. Note that the president of the I-Club at the time was Vartan Gregorian, a young Armenian born in Iran who later, like Guy, also got his PhD at Stanford, and became well known as an educator and an author of several books.

The friendship between Guy and I started immediately and went on for 67 years. He was then living in his Portola Valley house with his first wife, Frances, and his three children, and I lived as a bachelor in Palo Alto. Of course our relationship was interrupted when Guy went to work back East or in Paris while I pursued my fifty-year career at Stanford. Eventually, Guy became a professor at UC Berkeley, and married Karen around the time I married Gilda and moved to Atherton.

How can I summarize our friendship of 67 years? Great conversations, great French cheese, great common friends, great sunsets over the Golden Gate from their window, and very few disagreements, except that I could never convince Guy to come to the SF Opera with me: "Why should I sit for three hours watching and listening to people gargle in front of me?"

I always admired Guy's passion for painting and art restoration, and invariably enjoyed visiting their private museum. Lunches and dinners at 150 Montrose, prepared in artistic and culinary collaboration with Karen, were always memorable. Just as good as Chez Panisse!

With our backgrounds, Guy and I could never be nationalists. Yet I must admit that we both worshipped the moment in the movie Casablanca when the German officers stormed Rick's Café, and everybody stood up to sing La Marseillaise! Allons enfants de la patrie!

I miss Guy very much!

Memories of My Friend Richard Pantell

Richard and I met 70 years ago in 1955 at the Electronics Research Lab at Stanford.

Richard had just gotten his PhD in EE and I was embarking on mine.

We soon became friends and I got to know him and his wife Leona very well over time.

Two things struck me from the beginning about Richard: his scientific curiosity and how much his political opinions, which were very liberal, had been shaped by the Depression which he and his family had experienced when he was a kid. He was always very compassionate.

ESSAYS

Around 1958, we started a hobby together with a third friend, Jorge Fontana: Studying Quantum Mechanics. It was a lot of fun and quite an effort.

As those who hnew him, Richard was a very smart and eclectic scientist. He delved into many daring and diverse research fields.

It was not much of a surprise that by 1969 he co-authored a book called "Introduction to Quantum Electronics." Five years later, around 1974, during a sabbatical in Australia, Richard prophetically became interested in another subject, environmental problems, and co-authored another book on Environmental Modeling.

Skipping our subsequent careers, his at Stanford and mine at SLAC, let's move into this century. By then, Richard had unfortunately lost his wife Leona. He later met his wonderful friend Carol, and together they moved to the Bradford Retirement Community where he organized excellent events and lectures.

About ten years ago, Richard and I got into the habit of meeting for lunch, about once a month or so. There was never a pre-planned theme to these luncheons but they were invariably most enjoyable, spanning travels, politics, science and everything else.

Our conversations were never trivial, and I am already missing them very much. I would omit something if I didn't tell you that our discussions became increasingly animated after the first election of President Trump.

Many of you may have attended Richard's momentous 90th birthday.

Just before his 91st, he shared with me that he had to renew his driver's license and take the test. As you might expect, he passed the test, good for five years. He confessed to me he only got one answer wrong. Oh well!

Richard seemed to be in great and lively shape until recently. To finish, I want to share something very special about his psychological make-up.

Sometime last February, during one of our lunches, we talked about our respective attitudes toward death. I told Richard that it seems very difficult for me to accept this idea of disappearing from the scene. He told me he had no such problem, and when I looked surprised, he answered: "Greg, I only worry about things that I can do something about. I have had a very good and full life, and there is nothing I can do about death." Period!

I was very impressed by this answer. Richard seemed to worry only about things he could do something about. I don't know how he managed that, but he obviously left us in that spirit, with great dignity.

We will miss him very much.

The Significance of the 1967 Line

Today, by courtesy of Americans for Peace Now, I was privileged to participate in an hour-long telephone conversation with Mr. Akiva Eldar, the chief political columnist for Haaretz in Israel. Because of unusual circumstances, there were only six or seven people on the line and it was possible to have a very informative back-and-forth conversation. I will try to summarize the contents of this conversation below.

One of the first interesting things Mr. Eldar said is that for about two or three weeks, with Netanyahu's knowledge if not his consent, Shimon Peres and Mahmoud Abbas have been holding confidential talks to explore if there is any way for the Palestinians to return to the negotiating table, short of an openly declared Israeli moratorium on the settlements, which is extremely unlikely.

The window of opportunity for a resumption of talks before this past week was already extremely narrow, and Netanyahu's list of five or six pre-conditions spelled out in his speeches before AIPAC and the Congress closed this window even more, despite all the applause and standing ovations.

It is for this reason that it is important to understand the significance of the June 4th, 1967 line to the Palestinians. Neither President Obama, nor Israel, nor the Palestinians believe that there will be an exact return to the 1967 line if and when peace negotiations succeed. What the 1967 line did delineate and imply for the parties since the Oslo Agreement in

1993 is that it established the ultimate proportions of land to be controlled by each side: 78% for Israel and 22% for Palestine. Any swaps on either side of the line, as referred to by President Obama, would have to preserve these ratios, when taking the sum of all the swaps into account.

Hence, if the Palestinians could be persuaded that this rule will be fully respected in any final agreement, there is a small probability that they might feel more inclined to return to the table without a moratorium on settlements. This is a card that Obama and Peres might play when they talk to Mahmoud Abbas.

It is important to realize that timing now to getting back to the table is of the essence. In September, if nothing constructive happens in the mean time, the Palestinians will push for statehood before the UN. In the General Assembly where there is no veto, they will almost certainly get a majority and prevail. Eldar pointed out that this will be a symbolic victory but will probably not change real conditions on the ground, although it may trigger violence and repression. If they go to the Security Council, the US and perhaps one other country will veto the request and block the official legal creation of a Palestinian State. This will result in an ambiguous and unstable situation. It would be much better if this confrontation could be avoided ahead of time.

Hence, a reference to the 1967 line is an important starting point, and President Obama was right to insist on it.

Peace Between Israel and the Palestinian Authority is the Only Solution

Hamas' horrendous attack on Israel on October 7th, 2023, was its attempt to make a peaceful settlement in the area impossible. We can't let it succeed. Violence on Israel and the Palestinians has not worked for years and is not the solution. Antisemitism is not the solution. Contempt for the Palestinians is not the solution. Attacking Iran does not help. The two-state solution, long blocked by the Netanyahu government, is still the only way out of this tragedy.

The Oslo accords of 1993 and 1995 marked a huge step toward peace in that Israel and the Palestine Liberation

Organization (PLO) recognized their mutual rights to exist and to carry forward a peace process between them. The prospects looked promising if it hadn't been for the assassination of PM Itzak Rabin in November 1995, and the election of PM Netanyahu in 1996. Ever since, the peace process launched by the Oslo accords has been thwarted by political maneuvers and the illegal expansion of the unbridled Israeli settlements on the West Bank. Lack of progress led to the parliamentary Hamas victory in Gaza in 2006 with its position that "Israel does not have the right to exist." This was a complete reversal of the conditions gained by the Oslo accords.

The endless Israeli campaign that by now has killed more than 60,000 people in Gaza, must stop. At this point, it seems like Netanyahu will not do this unless the US shows the courage to stop supplying him with arms to bomb Gaza. Might President Trump be convinced to support such a move?

GREGORY A. LOEW

The most effective way Israel can combat the Hamas ideology is by restarting the peace process with the Palestinians in the West Bank. Such a step is very difficult in the current tense atmosphere, but it is the only hope for peace. For this to happen, strong pressure with sticks and carrots from the UN. Recognition of a Palestinian state is only a first step. They must deal with the principal issues of contention between the parties: mutual land allocations consistent with the 1967 line and UN Resolution 242, a Palestinian capital in East Jerusalem, and mutual security arrangements. Negotiations should preferably be held in a neutral country like Switzerland and will need a skilled full-time mediator to move the process along swiftly. Somebody like George Mitchell in Ireland would be ideal.

It Is Time to Reform the UN Security Council

When the U.N. was founded by 50 countries in San Francisco in 1945, its principal mission was to protect the world from the scourge of war. For this, central responsibility was assigned to the Security Council (UNSC) to resolve international conflicts peacefully through negotiation and to use collective military force only as a last resort. Unfortunately, in the past 80 years, the UNSC has not lived up to its expectations. The purpose of this article is to analyze why this has happened and to come up with a viable plan to remedy the situation. Admittedly, this will not be easy.

The Cold War started almost immediately after WW2 and, in 1962, came close to blowing up the world with the Cuban missile crisis. The post-partition wars between India and Pakistan over Kashmir resumed several times. The Korean war that started in 1950 never really ended. The Vietnam war went on for almost fourteen years. Following the 1948, 1956, 1967 and 1973 wars with Arab states, violent conflicts between Israel and the Palestinians continue to this day. The Falkland war erupted in 1982. Iraq's Saddam Hussein, after his war with Iran, invaded Kuwait. There followed 9/11, the wars in Afghanistan, Iraq, Syria. Isis, and the Taliban. On top of this, we now have Russia's unprovoked war on Ukraine, and the war between Hamas, Israel and Iran. And amid all this tension, the world currently has nine nuclear states armed with about 12,400 warheads.

Why has the UNSC been so unsuccessful in averting most of these conflicts? There seems to be one dominant reason. In 1945, the consensus was that the UNSC should include five permanent members, i.e. the USSR, the USA, China, the UK and France (the victors of WW2) plus six non-permanent members (increased to ten in 1965), rotating every two years. The UK with its Commonwealth, France with its colonies, the USSR with all its republics, China and the US at the time represented a large fraction of the world's population, which is not the case today.

Enshrined in Articles 23 and 27 of the U.N. Charter, was the rule that any non-procedural resolution submitted to the Council had to be supported by 9 out of 15 countries to pass, and that, without exception, all five permanent members had to be included in this majority of nine, unless they abstained as the U.S. and Russia did on December 22nd 2023. This rule, without explicitly spelling out the word veto, gave them this effective power to block any action to which they objected, including military force. This, they have now done at least 266 times, to get their way: the US about 80 times, Russia about 120 times, China, the UK and France the remaining 60 times.

Can Articles 23 and 27 be modified to make the UNSC more effective? In the last few decades, endless attempts made by Secretaries General, various groups of countries in the General Assembly, national leaders, scholars and journalists, have failed. Given the power these Articles confer in their current form to the five permanent members, this is not surprising.

One possible modification of Articles 23 and 27 that I think may be viable because it reflects the geopolitical changes of the last 80 years, would be to add eight new members to the five current permanent members: these might include a balanced mix of countries like India, Pakistan, Nigeria, Egypt, Brazil, Japan, Germany and Italy or Mexico. By reducing the rotating members to eight, there would be a total of 21 countries. To pass, a resolution would need eight out of the thirteen permanent members plus six non-permanent members (14 in total), i.e. a 2/3 majority. This expansion would not be risk-free but would avoid the current paralysis of the UNSC. It could be enacted without touching Articles 41 through 51 that in their current form already authorize the Council to mobilize collective military action when needed.

Note also that Chapter 18, Article 109, part 1 spells out clearly that "A General Conference of the Members of the U.N. for the purpose of reviewing the present Charter may be held at a date and place to be fixed by a two-thirds vote of the members of the General Assembly…" Actual ratification of proposed amendments would also require a two-thirds majority vote, but would have to include all current five permanent members, a taller order that will take some time to negotiate. Such negotiations would be facilitated if the UNSC trained and made available skilled mediators like George Mitchell, architect of the Northern Ireland peace agreement.

Indeed, the plan outlined here (or some variation thereof) will not be realized overnight, nor will it solve all U.N. problems. But considering that the world currently spends close to $2 Trillion/year on defense, it is high time to find a more productive way to promote international peace and security.

The Essence of the U.S. Federal Social Security System 1935-2025

U.S. Social Security is one of the crown-jewels of the American safety net. It was enacted by President Franklin D. Roosevelt when he signed the Social Security Act of 1935. It came after a long chain of events and traditions starting with the Greeks, the Middle Age guilds, labor unions, pensions, the Great Depression, and so on.

Social Security is made up of two insurance programs, Old-Age and Survivors Insurance (OASI) and Disability Insurance (DI) which, for simplicity, are lumped together as one program (OASDI) with one Trust Fund. In a nutshell, this Trust Fund receives money from payroll taxes contributed by employers and employees, and pays it out to beneficiaries, according to established rules.

In 1983 the Greenspan Commission, realizing that the system needed more reserves, increased the Social Security payroll tax rates to a total of 12.4% (6.2 % for employers and 6.2% for employees) and every year until 2009, OASDI raised more money from these taxes than it pay out to beneficiaries: it consequently accumulated annual surpluses. Unfortunately, all administrations during this period borrowed these surpluses to pay for other on-going programs, several wars or tax cuts, whereas they could have used them to pay down the National Debt. To account for these loans, they deposited non-marketable Treasury Bonds as IOUs equal in value to the surpluses in the Trust Fund. These bonds bore annual interest like any other Treasury Bonds held by the public (domestic and foreign) and this interest was added to the moneys accumulated by the Trust.

The George W. Bush administration tried to privatize Social Security several times but fortunately did not succeed.

In 2009, funds raised and disbursed by OASDI were essentially equal, amounting to roughly $690B. The payroll contributions were made by 156 million people and their employers (12.4% on incomes now up to ~$107K), and they were paid out to 52 million beneficiaries, exactly a ratio of 3-to-1. The total Trust Fund accumulation in IOU bonds was about $2.5 Trillion which in that year earned an extra $118B in interest.

Now let us move to the present. At the end of 2024, the value of the Trust Fund was about $2.7 Trillion, and total outlays to beneficiaries were $1.45 Trillion. The 12.4% payroll contribution now up to an income cap of 176 thousand dollars/year raises about $1.2 Trillion. At this rate, the

Trustees of the Trust Fund predict that the fund will be depleted by 2033 and outlays will drop to about 75% of current values, unless reforms are made by then: small increases in the payroll tax rates and/or taxable cap, or other tweak to the law. None of these are unsurmountable if the political will is there.

Hallmarks of Totalitarian States

1) Countries where power is concentrated in a single leader and/or group

2) The group's leadership can be political, ideological, ethnic, economic, religious, military, or several of the above

3) The rule of law is replaced by the rule by law or decree.

4) Separation of church and state may be abolished

5) Executive, legislative and judicial power may be centralized under single leadership while the people lose their power

6) Elections may be suppressed

7) The press may be highly or totally controlled

8) Labor unions may be forbidden

9) Violent repression by secret police may be used against innocent victims

10) The government may be racist and/or xenophobic

11) Discrimination may exist against women, LGBTQ, other minorities

12) The government is free to lie as much as it wants

13) The truth is replaced by political propaganda

FY2025 and FY2024 Federal Budgets

The two budgets for 2025 and 2024 are shown at right for comparison. Nobody knows exactly yet what will happen with Trump's FY 2025 budget, except that he signed it. There are uncertainties in the GDP growth because of the onset of tariffs, the effect of tax cuts on total receipts, and the cuts in Medicaid and SNAP. The numbers will be quite different in FY2026 and 2027.

Federal Budgets Listed by Agencies	FY2025 President Trump Approved B$	FY2024. President Biden Proposal B$
Projected GDP	29.5	27,238
Totals Receipts in B$ (as % of GDP)	5485 (18.7)	5036 (18.5%)
Total Outlays in B$ (as % of GDP)	7266 (24.8)	6883 (25.3%)
Deficits in B$ (as % of GDP)	1781 (6.1)	1846 (6.8%)
1. Discretionary Programs	1940.5	1897
Security and Associated Agencies, Total	*1228.7*	1212
Defense	850	842
Overseas Operations		
Homeland Security	62.2	60.4
Veterans Affairs	153.7	137.9
State and International Affairs	64.4	70.5
NNSA	25	20
National Intelligence Program	73.4	101.2
Non security Agencies, Total	*711.8*	685
Agriculture	29.2	30.1
Commerce	11.4	12.3
Education	82	90
Energy (Non NNSA)	51	32
HHS	130.7	144
HUD	72.6	43.3
Interior	17.8	18.8
Justice	37.8	39.7
Labor	13.9	15.1
Transportation	25.4	27.8
Treasury	14.4	16.3
EPA	11	12
NASA	25.4	27.2
NSF	10.2	11.3
All Other Agencies + Programs	~179	165
1. Mandatory Programs	4400.5	4192
Social Security (OASI) and (DI)	1620	1459 [1209 from payroll taxes + 250 from SSFund
Medicare (HI) and (SMI, Parts B and D)	1201]	842 [369 from payroll taxes +473 fr.MTFund]
Medicaid and other health services	Not yet known	556

The Nuclear Weapons Age at 80 [1945-2025]

2025 marks the 80th anniversary of the first atomic bomb test in the US in New Mexico in July 1945 and the dropping of two US nuclear bombs respectively on Hiroshima and Nagasaki in August 1945. Shortly thereafter, the Cold War started, and the international nuclear arms race took off. In my book, *The Human Condition, Reality, Science and History*, I devoted a section to the international balance of terror regime that came along in these eighty years, which I am summarizing and updating here.

If you want to be up to date on the

status of arms control, please subscribe to Arms Control Today published by the Arms Control Association (ACA) based in Washington DC and led by Daryl Kimball.

In October 1962, the Cuban Missile Crisis came very close to blowing up the world. By 1985, the arms race had increased the world's arsenals to close to 70,000 nuclear warheads with yields growing from tens of kilotons to tens of megatons. Fortunately, over this period, leaders gradually accepted the "MAD" principle of Mutual Assured Destruction. This means that nuclear weapons serve only as mutual deterrents but cannot help to win a war if both sides possess them. This realization led to various arms control reduction treaties like the Non-Proliferation Treaty (NPT), the Comprehensive Test Ban Treaty (CTBT) and the Strategic Arms Reduction Treaty (START I) between the US and Russia. As a result, by the beginning of 2025, there were only about ~12,400 warheads worldwide, as shown below:

Russia: 5580 warheads, with 4380 in the military stockpile, including 1549 strategic warheads deployed on delivery vehicles and up to 2000 tactical weapons, and additional 1200 warheads awaiting dismantlement.

US: 5225 warheads, with 3748 in the military stockpile, including 1419 strategic warheads deployed on delivery vehicles and about 100 tactical weapons, and additional 1477 warheads awaiting dismantlement.

China: 600 warheads

France: 290 warheads

UK: 225 warheads

India: 172 warheads

Pakistan: 170 warheads

Israel: 90 warheads

North Korea: 50 warheads

Despite these reductions, a fraction of

these arsenals if used would still be sufficient to make the earth uninhabitable, except perhaps for cockroaches. The current New START Treaty will expire by February 2026. Agreeing to do this with lower caps would be highly desirable but not obvious with the current relations between the US and Russia around the war in Ukraine, the situation between China and Taiwan and the tensions in the Middle East.

Actually, as long as any nation possesses any nuclear weapons, the risk of a deliberate or accidental launch, or a terrorist act is just too high. Nothing short of the elimination of all nuclear warheads can protect humanity. This is the goal of the International Campaign to Abolish Nuclear Weapons (ICAN) to support the UN Treaty to ban nuclear weapons or the Treaty on the Prohibition of nuclear weapons. The treaty is currently supported by 73 states parties and ICAN is a coalition of over 650 partner organizations including international non-governmental organizations (NGOs). They deserve our support!!!

A small number of nuclear weapons should be turned over to a special branch of the UN in case they are needed to deflect the trajectory of an asteroid threatening the earth.

Let's face it: can our species HOMO become SAPIENS enough to take this step? Do we really want to relegate our earth to cockroaches?

Adolescence, Puberty and Adulthood

Adolescence is a process that starts arbitrarily around the age of 10 or 12 and continues for a decade or so until adulthood. Puberty used to start around the age of 13 but is currently starting earlier in girls. After puberty the child can begin to reproduce sexually: it is a one-way transition. Around this time, the individuals have generally become conscious of their sexual orientation, straight or LGBTQ. Just under 10% of children become LGBTQ, for reasons that are not well understood.

Adolescence is a much more gradual process with a sequence of mental development phases. These phases have to do with relationships (typically some distancing from parents and formation of new bonds with peers), education, the influence of social media, choice of profession, cultural environment and other factors like political involvement and voting. As of lately, there is some evidence that some adolescents have more difficulties coping with societal changes and tend to be more easily depressed. Some adolescents never grow up completely. What is clear is that there is too long a gap between puberty and maturity to deal with it.

A platonic relationship is generally easy to handle; a sexual relationship is a little harder; teen-age pregnancy is a disaster; the preparation to enter a long relationship and/or a strong marriage is the ultimate challenge.

Why Human Colonies on Mars Are a Hopeless Project

1) Getting humans to Mars in about 8 months is only a small part of the problem. Cosmic radiation impinging on Mars is about 50 times more intense than on Earth. Humans would have to spend most of their time in deep shielded tunnels to survive.

2) Mars is much smaller than Earth and its gravity is only 38%. Human bones would gradually degrade

3) Diurnal temperature fluctuations on Mars are about 130 degrees F

4) Humans would have to have their

own sources of energy, presumably solar panels, nuclear reactors and batteries, and keep them going for the duration.

5) Humans could not grow food.

6) There is no atmospheric oxygen to breathe on Mars, and insufficient water, as far as we know.

7) Colonies would have to start with several men and women, all highly educated, doctors, engineers, etc.

8) Women would have to be able to conceive, give birth and raise healthy children, something not obviously possible in that environment

9) How about eating, urinating and defecating?

10) How much of their time would all settlers have to spend inside their space-suits?

11) Would they ever have the means to return to Earth?

12) Part of the time, the sun is between Mars and Earth!

13) Do I hear any volunteers?

Souvenirs d'Evian, Août 1939

Pour le 31 mai 2019, j'avais réservé une chambre à l'Hôtel Parc 55 à San Francisco pour les deux jours de la Convention Démocrate annuelle de Californie. En arrivant dans ma chambre, j'avais découvert sur la table deux bouteilles d'eau d'Evian. Voyant le prix de $6 par bouteille, je m'étais dit que je m'abstiendrais de les ouvrir. Après m'être enregistré à la Convention au Moscone Center, j'ai diné à l'hôtel où la nourriture fût très salée. Pendant la nuit, je me suis réveillé avec une forte crampe à la jambe. Sur quoi, ne sachant pas quoi faire, j'ai décidé qu'après tout, une bouteille d'Evian pourrait me soulager (en diluant le sel dans mon système!), même au prix de $6! Aussitôt pensé, aussitôt fait. La crampe se calma.

Ce que je n'avais pas prévu, c'est qu'au milieu de cette nuit, j'allais me souvenir de nos vacances à Evian, il y a presque quatre-vingts ans, comme dans un film.

Depuis Novembre 1938, nous habitions dans notre joli appartement au 5eme étage du 179 rue de la Pompe, dans le 16eme arrondissement. Tout était blanc-crème et sentait encore la peinture fraiche. Du balcon, on apercevait l'Avenue du Bois où avaient défilé le Roi Georges VI et la Reine d'Angleterre lors de leur visite à Paris, en voiture ouverte en compagnie du Président Lebrun et de Madame Lebrun, suivis du régiment magnifique des Spahis. Cet été, nous allions passer un mois de vacances en Savoie où nous avions loué une maison à Evian, sur les bords du Lac Leman.

Je ne me souviens pas bien de la maison, mais de la rotonde vitrée où se trouvait la salle à manger, on voyait le lac et Lausanne, en face. Comme d'habitude à l'époque en France, le linge n'était pas inclus dans la location, et nous étions arrivés avec deux énormes valises bleues pleines de serviettes et de linge à lit. Il faisait très beau. Je crois que notre nouvelle nounou, Marcelle Bardin, et notre femme de chambre, Suzanne Petit, étaient venues avec nous. Sébastien était né le 6 juin (cinq ans avant D-Day), avait deux mois et passait son temps dans son berceau. Grand-mère Anna de Bucarest était aussi venue, comme elle le faisait plus ou moins tous les étés. Elle nous adorait. Elle était maigre comme un clou et lors de chaque repas, Papa l'engueulait et tachait de la faire manger. Il criait en Allemand: "Mama, iss!"

Nos journées se passaient en petites promenades. Je me rappelle qu'un jour nous étions allés visiter la source d'Evian où l'on mettait l'eau en bouteille, à la chaine. C'était très intéressant. A part cela, j'avais une espèce de routine qui commençait agréablement dans la matinée avec une leçon de gym et de course pas loin de la plage, avec un prof plutôt sympathique. J'étais très fort en course et je gagnais tout le temps. On me demandait comment je faisais, et je disais aux autres gosses que je prenais "un élan crétin." Malheureusement, la matinée se terminait toujours mal parce que Papa arrivait à la plage, et avec le même prof, me faisait plonger à partir d'un tremplin, J'avais la terreur de mettre la tête sous l'eau depuis ma plus tendre enfance. Et les choses ne s'amélioraient pas.

L'après-midi, Monique et moi jouions au jardin. Derrière notre maison, il y avait une petite dépendance où habitait la gardienne avec sa fille. Il y avait aussi un poulailler. La fille, d'environ 5 ou 6 ans, était très amusante et jouait avec nous à cache-cache et à la marelle. Je me souviens qu'un jour elle dit à Monique et moi qu'elle voulait nous montrer comment elle faisait caca. Sur ce, elle a levé sa robe, baissé sa culotte et comme prévenu, tout heureuse, elle a fait un petit caca devant nous. Nous étions assez épatés.

Derrière Evian se hisse une montagne assez raide, la Dent d'Oche. Evidemment elle n'avait pas échappé à Papa qui, inévitablement, avait décidé de l'escalader. J'avais 9 ans et il avait très envie que j'aille avec lui, ce qui fût fait bien que Maman ne fût pas enthousiaste. Il fallait monter au refuge dans la journée et y passer la nuit. Tout s'était bien passé mais le lendemain matin, quand je me suis réveillé au dortoir, Papa n'était pas à côté de moi (il était parti seul au sommet sans me prévenir) et je me suis mis à pleurer. Une dame est venue me consoler en me disant que mon papa reviendrait bientôt. Heureusement, il arriva peu après avoir conquis la Dent tout seul!

L'atmosphère politique, au cours du

mois, s'assombrissait rapidement. Papa, inquiet, lisait tout le temps les journaux, L'Œuvre et le Temps. Nous n'avions bien sûr pas de TSF ni de voiture. On sentait que les choses allaient de plus en plus mal en Europe. Grand-mère, par précaution, fut renvoyée à Bucarest. Le 23 août, on apprit que Molotov et Ribbentrop avaient signé un pacte de non-agression entre l'URSS et l'Allemagne, chose dont je ne devais pas comprendre l'importance mais qui était de très mauvais augure. A partir de ce jour, la Défense Anti-aérienne française avait décrété le couvre-feu sur tout le territoire. Je me souviens combien c'était bizarre d'être dans l'obscurité tous les soirs, dehors, alors que la côte suisse en face brillait de lumière.

Finalement, nous quittâmes Evian en train de nuit pour Paris le 31 août. Le matin du 1er septembre, nous étions de retour à la rue de la Pompe. Je me souviens comme hier que Monsieur Gross (fondé de pouvoir de mon père) vint nous rendre visite. Tout avait encore l'air calme et Monsieur Gross nous quitta avant midi. A peine descendu dans la rue, il trouva un journal au stand d'en face avec une immense manchette: "Hitler envahit la Pologne." Cinq minutes après il était de retour chez nous avec la nouvelle. Peu après, on apprenait que toutes les familles avec petits enfants devaient quitter Paris le plus vite possible. C'était le premier exode, celui de 1939!

Quelques heures plus tard, nous étions dans deux taxis, avec toutes nos valises non-déballées et Suzanne Petit, en route vers la Gare de l'Ouest. Monique pleurait et demandait à Maman pourquoi les gens font la guerre. Arrivés à la Gare de l'Ouest, il y avait partout des queues interminables devant les guichets. Papa nous fit nous asseoir sur les grandes valises bleues et se mit en queue. Vers quatre heures de l'après-midi, il avait finalement obtenu des billets pour Rennes. Le train devait passer par Le Mans, le patelin de

Suzanne Petit où elle voulut descendre. Nous perdîmes totalement contact avec elle, mais…pas pour toujours! Maman la retrouva tout à fait par hasard comme femme de chambre à l'hôtel St Philippe du Roule vers 1965!

Vers le soir, nous arrivions à Rennes et trouvions deux chambres d'hôtel sur la place principale. Deux jours après, le 3 septembre, la France et la Grande Bretagne déclaraient la guerre à l'Allemagne en vertu d'un ancien pacte avec la Pologne. La Deuxième Guerre Mondiale avait commencé.

Le reste est une autre histoire.

Cela dit, l'eau d'Evian est bonne, et finalement les 6 dollars ne sont pas apparus sur ma note d'hôtel.

Au Clair de la Lune

When brought up in the French culture as my siblings and I were, learning to sing children's songs was part of the tradition. Our father used to accompany us on the piano. One of the most well-known songs was "Au clair de la Lune." This song consisted of four parts of which, for reasons you the reader will find out with a little French, we as kids learned only the first two:

> Au clair de la Lune
> Mon ami Pierrot
> Prête-moi ta plume
> Pour écrire un mot
> Ma chandelle est morte
> Je n'ai plus de feu
> Ouvre-moi ta porte
> Pour l'amour de Dieu

GREGORY A. LOEW

Au clair de la Lune
Pierrot répondit
Je n'ai pas de plume
Je suis dans mon lit
Va chez la voisine
Je crois qu'elle y est
Car dans sa cuisine
On bat le briquet

Au clair de la Lune
L'aimable Lubin
Frappe chez la brune
Qui répond soudain
Qui frappe de la sorte
Il dit à son tour
Ouvrez-moi la porte
Pour le Dieu d'Amour

Au clair de la Lune
On n'y voit qu'un peu
On cherche la plume
On cherche le feu
En cherchant d'la sorte
Je n'sais c'qu'on trouvera
Mais je sais qu'la porte
Sur eux se ferma

Acknowledgements

I wish to thank my son George E. Loew: who encouraged me to publish this collection of essays, saying that the main criterion for doing this should be if it gave me pleasure, which it does.

I also want thank my daughters Linda Maki and Florence Surratt, my good friend Ben Lenail for their encouragement and help, and my housekeeper Bertha Perez for her faithful support for 50 years.

Finally, given my arthritic disabilities, these acknowledgements would not be complete without recognizing the help of three excellent caregivers of Ethiopian origin: Serkalem Seda, Emebet Kebede and Aynalem Kebede. The latter two, in addition to their caregiving, were extremely helpful to me with their computer skills.

www.ingramcontent.com/pod-product-compliance
Lightning Source LLC
Chambersburg PA
CBHW060530080526
44586CB00012B/690